Zip

written by Gale Clifford
illustrated by Shelley Dieterichs

SAXON PUBLISHERS

THIS BOOK IS THE PROPERTY OF:

STATE_____	Book No. _____
PROVINCE_____	Enter information
COUNTY_____	in spaces
PARISH_____	to the left as
SCHOOL DISTRICT___	instructed
OTHER_____	

ISSUED TO	Year Used	CONDITION	
		ISSUED	RETURNED

PUPILS to whom this textbook is issued must not write on any page or mark any part of it in any way, consumable textbooks excepted.

1. Teachers should see that the pupil's name is clearly written in ink in the spaces above in every book issued.
2. The following terms should be used in recording the condition of the book: New; Good; Fair; Poor; Bad.

Zip is last.

Zip must run fast.

Zig, zag, zig, zag.

Zip must not lag.

Zip is hot, hot, hot!

Zip must stop.

A bug is not fun.

Run, run, Zip! Run!

The End

Understanding the Story

Questions are to be read aloud by a teacher or parent.

1. What is the title of this book?

2. What kind of animal is Zip?

3. What sound does the bug make?

4. Why does Zip run away from the bug?

Answers: 1. Zip 2. a zebra 3. buzz, buzz, buzz 4. Possible answer: because the bug is bothering him

---

Saxon Publishers, Inc.
Editorial: Barbara Place, Julie Webster, Grey Allman, Elisha Mayer
Production: Angela Johnson, Carrie Brown, Cristi Henderson

Brown Publishing Network, Inc.
Editorial: Marie Brown, Gale Clifford, Maryann Dobeck
Art/Design: Trelawney Goodell, Camille Venti, Andrea Golden
Production: Joseph Hinckley

© Saxon Publishers, Inc., and Lorna Simmons

All rights reserved. No part of the material protected by this copyright may be reproduced or utilized in any form or by any means, in whole or in part, without permission in writing from the copyright owner. Requests for permission should be mailed to: Copyright Permissions, Harcourt Achieve Inc., P.O. Box 27010, Austin, Texas 78755.

Published by Harcourt Achieve Inc.

Saxon is a trademark of Harcourt Achieve Inc.

Printed in the United States of America
ISBN: 1-56577-952-5

4 5 6 7 8 446 12 11 10 09 08 07 06

Saxon Phonics and Spelling K

Phonetic Concepts Practiced

z (Zip, buzz)

ISBN 1-56577-952-5

Grade K, Decodable Reader 6
First used in Lesson 67